What's Inside?

General Science

So Many Exciting Experiments

	PAGE
Getting Started	2

CORE EXPERIMENTS

	PAGE
Discovering Magnetic North	8
The Strength of a Magnetic Field	9
Testing Materials for Magnetic Attraction	10
Magnetic Lines of Force	11
Floating Magnets	12
Magnetic Strength	13
Challenge	14
Levitating Paper Clip	15
Making a Compass	16
Make an Electromagnet	17
Making a Stronger Electromagnet	18
Changing the Volume of Air	19
Using a Tuning Fork to Demonstrate Wave Patterns	20
Vibration of Reeds	21
Make a Button Spinner	22
Color	23
Strange Colors	24
White Light	25
A Beautiful World	26
Making Rainbows	27
Reflection	28
Making a Periscope	29
The Kaleidoscope	30
Chromatography	31
Shadows	32

EXTENSION ACTIVITIES

	PAGE
Repulsion Soccer	33
Boat Races	34
Paper Maze	35
Pirate Treasure	36
Cattle Roundup	37
Transmission of Sound	38
Communication	39
Variations on a Button Spinner	40
Amplitude and Frequency	41
Fun with a Prism	42
Strange Lines of Light	43
Strange Chromatography	44
Seeing Around Corners	45
More Reflection	46
Multiple Images	47
Mirror Image Painting	48
Cleaning an Aquarium	49
Silent Signals	49
Use Shadows to Tell the Time	49
Didgeridoo	50
Ancient Musical Instrument	50
Use Our Knowledge of Reflection	50
Even More Reflections	51
Is This a Kaleidoscope?	51
Sounds Travel Better in Denser Air	51
Say What? Understanding Difficult Words	52
Notes	54

USE THE NOTES PAGES AT THE BACK OF THIS WORKBOOK TO RECORD YOUR ANSWERS.

GETTING STARTED — General Science

Note to parents:

Here's an exciting and creative way of introducing young children to the fascinating world of real science.

Each kit has been designed to provide a hands-on approach. The components of the kit equip your children to conduct a series of scientific experiments, allowing them to learn.

The easy, step-by-step instructions and the detailed diagrams in the workbook lead children in a developmental and educational way through 50 activities. By gathering household items that are readily available, children are able to conduct even more fun activities.

While science is often viewed as a difficult and uninteresting area of study, the purpose of these kits is to alter that misconception and make science fun-filled and enjoyable.

Experiments and extra activities encourage children to explore for themselves. Each kit promotes individual learning, investigative and problem-solving skills, and stimulates the inquiring mind.

Note to young scientists:

ALWAYS REMEMBER THAT EXPLORING SCIENCE IS AN EXCITING ADVENTURE, BUT YOU NEED TO HANDLE THE COMPONENTS AND ANY OTHER HOUSEHOLD ITEMS THAT YOU WILL BE USING IN THE EXPERIMENTS WITH GREAT CARE AND ATTENTION.

YOU SHOULD ALWAYS HAVE A PARENT OR ANOTHER RESPONSIBLE ADULT WITH YOU WHEN YOU DO THE EXPERIMENTS IN THIS WORKBOOK.

GETTING STARTED

General Science

What are all these things?

Let's begin by familiarizing ourselves with some of the components in your kit and some that you will have to find around the house. The following four pages contain images of the components and what they are called.

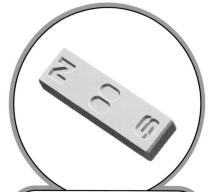
Bar magnet
WILL BE FOUND IN YOUR KIT

Ring magnets
WILL BE FOUND IN YOUR KIT

Iron filings
WILL BE FOUND IN YOUR KIT

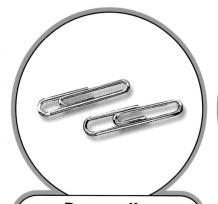
Paper clips
WILL BE FOUND IN YOUR KIT

Colored glasses template
WILL BE FOUND IN YOUR KIT

Cork
WILL BE FOUND IN YOUR KIT

Nail
WILL BE FOUND IN YOUR KIT

Needle
WILL BE FOUND IN YOUR KIT

Prism
WILL BE FOUND IN YOUR KIT

GETTING STARTED

General Science

Plastic tube
WILL BE FOUND IN YOUR KIT

Pan pipe
WILL BE FOUND IN YOUR KIT

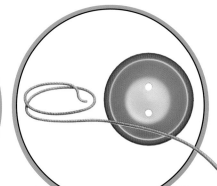
Button and cord
WILL BE FOUND IN YOUR KIT

Cardboard base & dowel
WILL BE FOUND IN YOUR KIT

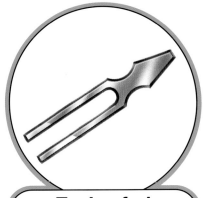
Tuning fork
WILL BE FOUND IN YOUR KIT

Straws
WILL BE FOUND IN YOUR KIT

Piece of foam
WILL BE FOUND IN YOUR KIT

Plastic safety mirror
WILL BE FOUND IN YOUR KIT

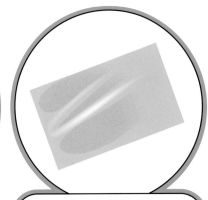
Colored cellophane
WILL BE FOUND IN YOUR KIT

GETTING STARTED

General Science

Shadow templates
WILL BE FOUND IN YOUR KIT

Color dot template
WILL BE FOUND IN YOUR KIT

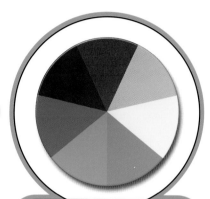
Color wheel
WILL BE FOUND IN YOUR KIT

Mirror templates
WILL BE FOUND IN YOUR KIT

Cotton thread
WILL BE FOUND IN YOUR KIT

Toothpick
WILL BE FOUND IN YOUR KIT

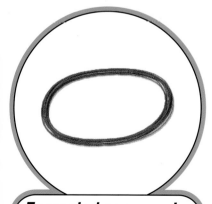
Enameled copper wire
WILL BE FOUND IN YOUR KIT

Battery holder
WILL BE FOUND IN YOUR KIT

Alligator clips
WILL BE FOUND IN YOUR KIT

Getting Started

General Science

AA battery
YOU WILL NEED TO FIND

Paper-towel strips
YOU WILL NEED TO FIND

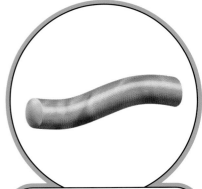
Modeling clay
YOU WILL NEED TO FIND

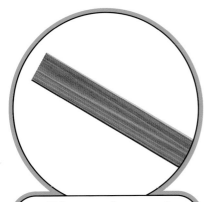
Piece of wood
YOU WILL NEED TO FIND

Adhesive tape
YOU WILL NEED TO FIND

Spoon
YOU WILL NEED TO FIND

Flashlight
YOU WILL NEED TO FIND

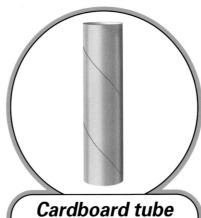
Cardboard tube
YOU WILL NEED TO FIND

Tin can
YOU WILL NEED TO FIND

General Science

Come into the lab and let the fun begin.

It's experiment time!

EXPERIMENT 1
Discovering Magnetic North

General Science

USE AN EXISTING PERMANENT MAGNET TO DETERMINE MAGNETIC NORTH AND THEREFORE OTHER COMPASS POINTS.

What you will need:
- 1 bar magnet and a piece of cotton thread

Here's what to do:
- Tie the thread securely to the center of the bar magnet.
- Suspend the magnet from the thread, by about 6 inches.
- Check that the magnet is balanced evenly at its center.
- Adjust the balance if necessary.
- Tie the thread to any place that allows the magnet to be suspended in midair, but well away from any other magnets or metallic objects.
- Notice that the magnet will move until it finally comes to a rest.
- This position will indicate that the magnet is aligned with the earth's magnetic field.
- On the magnet provided, an *N* on one end will indicate north and the *S* on the other end will indicate south.
- If you have a small magnetic compass you can check whether the direction is correct.
- The bar magnet is indicating the directions of the North and South Poles.

Where are east and west?

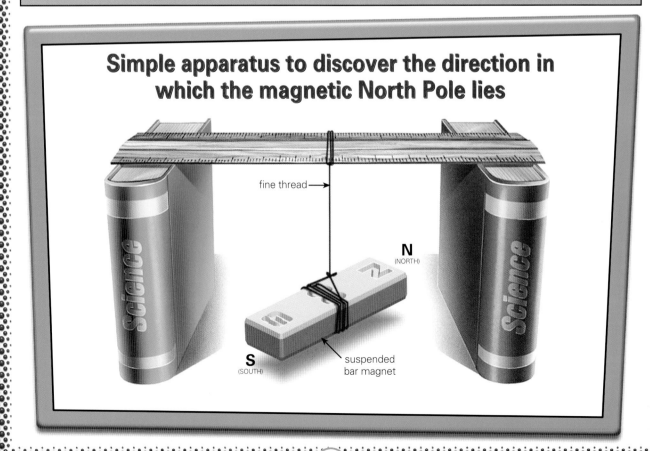

Simple apparatus to discover the direction in which the magnetic North Pole lies

EXPERIMENT 2
The Strength of a Magnetic Field

General Science

DETERMINE HOW MANY PAPER CLIPS A MAGNET CAN SUPPORT.

What you will need:

- 1 bar magnet and several paper clips

Here's what to do:

- Place the bar magnet over a desk or tabletop so that one end is in midair and balanced so that it will not topple over as you add paper clips to its end.
- Begin adding paper clips in a chain under the end of the bar magnet.
- Do not clip the paper clips together. They will cling together by magnetic attraction.
- Through this activity you can see that magnetism is transferred through the metal clips, each clip in effect becoming a temporary magnet.

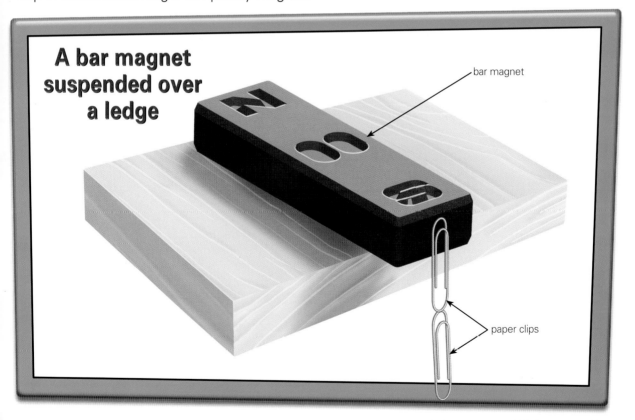

A bar magnet suspended over a ledge

How many paper clips does the magnet support?

- Try adding each paper clip to the top of the chain rather than the end.

How many clips can be joined in this manner?

EXPERIMENT 3
Testing Materials for Magnetic Attraction

General Science

HOW PERMANENT MAGNETS ATTRACT OR "STICK" TO A RANGE OF MATERIALS.

What you will need:

- Plastic tube, cardboard base, cork, a nail, enameled copper wire, piece of foam, and a sewing needle

Here's what to do:

- Use either end of the bar magnet to test each of the samples.
- Only some of the materials will be attracted to the magnet.
- Classify the materials into two groups: magnetic and nonmagnetic.
- Use your magnet to test other household items to see if they are magnetic or nonmagnetic.

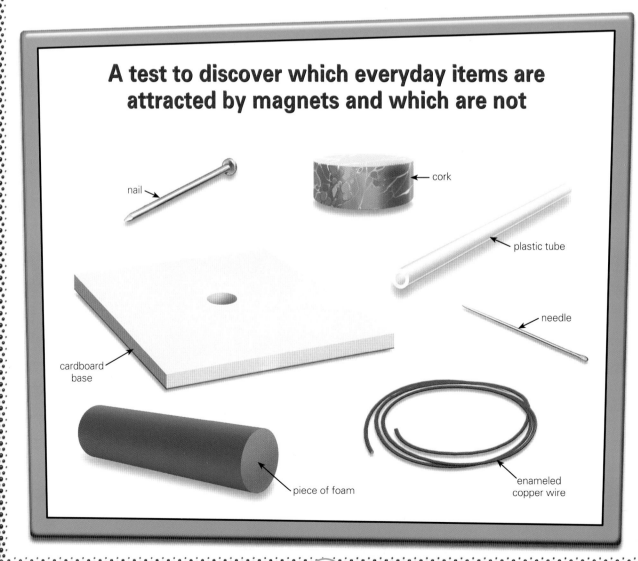

A test to discover which everyday items are attracted by magnets and which are not

EXPERIMENT 4
Magnetic Lines of Force

General Science

DEMONSTRATE LINES OF MAGNETIC FORCE.

What you will need:
- 1 bar magnet, 1 ring magnet, and a container of iron filings

Here's what to do:
- In previous activities, you demonstrated that a magnet will align itself with the earth's magnetic field and that magnets can both attract and repel themselves and some other materials.
- Shake the container of iron filings so that the filings are more or less evenly distributed within the container.
- Place a ring magnet under the base of the container. Gently tap the case.

What do you observe?
- Compare the difference in the pattern of iron filings when the bar magnet is placed under the container.

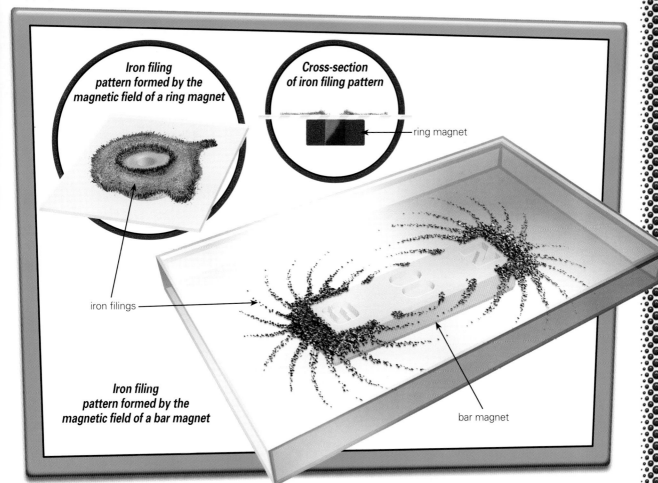

* As with the earth's invisible lines of magnetic force, magnets also have invisible lines of force extending out from each pole. The iron filings align themselves in this magnetic field, giving you a visible indication of their existence.

EXPERIMENT 5
Floating Magnets

General Science

SEE HOW PERMANENT MAGNETS HAVE FORCES OF ATTRACTION AND REPULSION.

What you will need:
- 3 ring magnets, a small dowel, and a cardboard base with a hole

Here's what to do:
- Press the dowel into the hole in the center of the cardboard base.
- Make sure that it is upright.
- Place a ring magnet over the dowel.
- Select a second ring magnet and slowly place it over the plastic dowel until the first magnet either jumps up to meet it or the second magnet floats (is repelled).
- Place all three magnets over the dowel so that they repel each other.
- Try placing two attracted magnets over the dowel. Place the third magnet so that it is repelled.

Does it float higher?

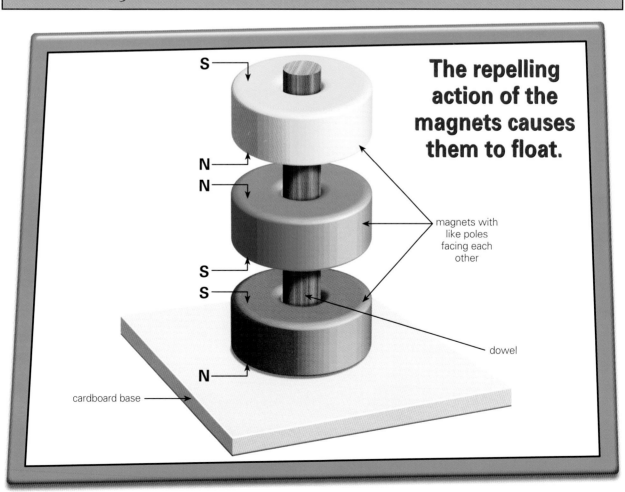

The repelling action of the magnets causes them to float.

EXPERIMENT 6
Magnetic Strength

General Science

TEST THE STRENGTH OF YOUR RING MAGNETS.

What you will need:

- 3 ring magnets

What you will need to find:

- Paper

Here's what to do:

- Place two ring magnets together so that they attract each other.

- Progressively insert sheets of paper between the magnets.

How many sheets of paper can be inserted before the magnets fail to attract each other?

- Repeat this using the third ring magnet.

Are all of the magnets of the same strength?

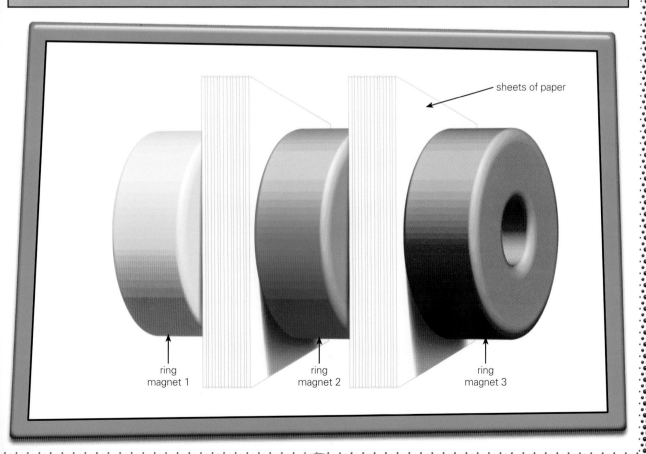

Experiment 7
Challenge

General Science

SEE IF YOUR FRIENDS CAN USE MAGNETISM TO SOLVE A PROBLEM.

What you will need:
- 1 bar magnet and 6 paper clips

What you will need to find:
- A glass of water

Here's what to do:
- Fill the glass with water.
- Place the six paper clips in the water.
- Challenge a friend to remove the paper clips without touching or emptying the glass.
- When your friend gives up, use your bar magnet to remove the paper clips by holding the magnet along the side of the glass to attract the paper clips. Then drag the magnet up the side of the glass to the top, where you can pick up the paper clips without touching the water.

EXPERIMENT 8
Levitating Paper Clip

General Science

OBSERVE THE STRENGTH OF A MAGNET.

What you will need:
- 1 paper clip, 1 bar magnet, and cotton thread

What you will need to find:
- Adhesive tape

Here's what to do:
- Tie a 12-in. piece of fine cotton thread to a paper clip.
- Tape the free end of the thread to a tabletop.
- Pick up the paper clip with the bar magnet and slowly move the magnet to tighten the thread.
- When the thread is taut, slowly move the magnet until it pulls clear of the paper clip but still attracts it enough to keep it suspended in air.

How big a gap can you get between the magnet and the paper clip before the paper clip falls? How many pages of a book can you put between the magnet and the paper clip before the strength of the magnet's field is reduced enough to make the paper clip fall?

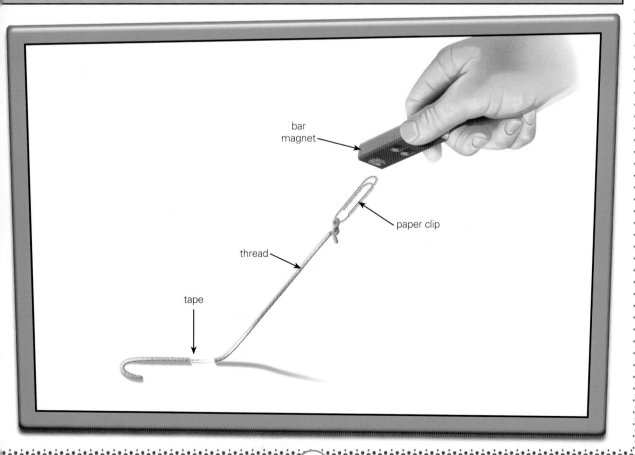

EXPERIMENT 9
Making a Compass

MAGNETIZE A PIN AND USE IT AS A SIMPLE COMPASS.

What you will need:
- 1 sewing needle, 1 bar magnet, and 1 cork

What you will need to find:
- A piece of adhesive tape and a bowl of water

Here's what to do:
- Place the needle on a flat surface and align it in a north-south direction.
- Use the south end of the bar magnet to stroke the needle about 20 times.
- Fill a bowl with water.
- Tape the needle centrally over the cork and then allow it to float in the bowl of water.
- Make sure that the bar magnet is well away from your experiment.
- The needle has become a small magnet and has its own north and south poles. The needle will align itself with the earth's magnetic field and will indicate north and south.

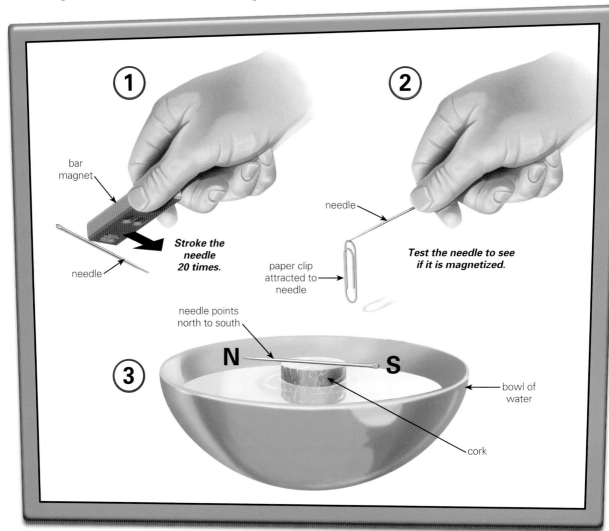

EXPERIMENT 10
Make an Electromagnet

General Science

USE THE POWER OF A BATTERY TO CREATE AN ELECTROMAGNET.

What you will need:
- 1 metal nail, the battery holder, 2 alligator clips, a length of enameled copper wire, and a paper clip

What you will need to find:
- 2 AA batteries and a piece of sandpaper

Here's what to do:
- Using the metal nail as a core, wind 20 turns of the enameled copper wire around the nail.
- Use sandpaper to remove enamel from each end of the enameled copper wire. A half-inch of enamel should be removed from each end of the wire.
- Connect your electromagnet into a circuit consisting of the battery pack connected to the other ends of the copper wire.
- When the circuit is complete, use the electromagnet to pick up a paper clip.
- Disconnect one copper wire from the circuit.

Will the electromagnet pick up a paper clip?

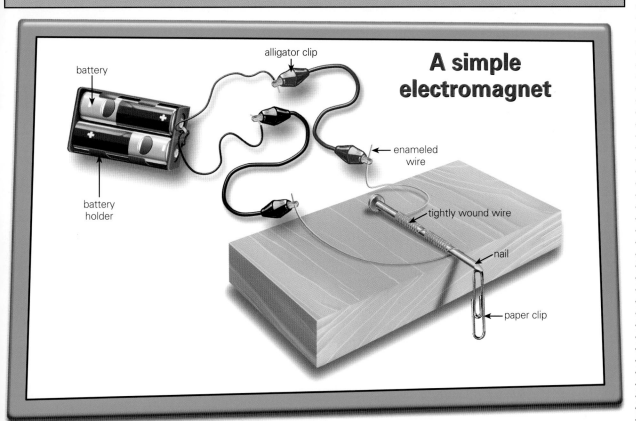

A simple electromagnet

✱ *You will notice that the electromagnet is not a permanent magnet like the bar magnet. It only works while in an active circuit.*

General Science

SEE IF YOU CAN MAKE A STRONGER ELECTROMAGNET.

What you will need:
- 1 metal nail, the battery holder, 2 alligator clips, a length of enameled copper wire, and paper clips

What you will need to find:
- 2 AA batteries and a piece of sandpaper

Here's what to do:
- See how many paper clips can be picked up by your original electromagnet.
- Make a stronger electromagnet by winding more of the enameled copper wire around the nail.
- Wind back over the coils you have already made until all the wire is used up.

How many paper clips will this electromagnet pick up?

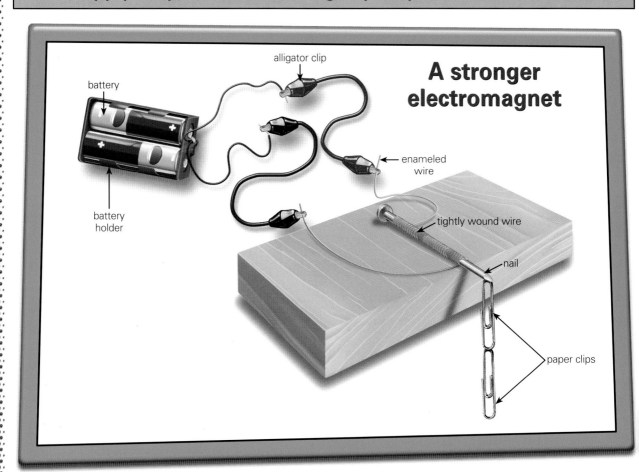

A stronger electromagnet

✱ *If the coils become too hot, disconnect the battery.*

EXPERIMENT 12
Changing the Volume of Air

General Science

DEMONSTRATE THE EFFECT OF CHANGING THE VOLUME OF AIR IN A WIND INSTRUMENT.

What you will need:
- The plastic tube fitted with a plunger (the Pan pipe)

Here's what to do:
- Hold the open end of the Pan pipe to your lips.
- Hold the end of the plunger using your other hand.
- Bring the opening next to your lips (but not over the tube) and then firmly blow into the Pan pipe.
- Listen to the sound created.
- While blowing again and creating another sound, gently push the plunger upward.

What happens to the sound?

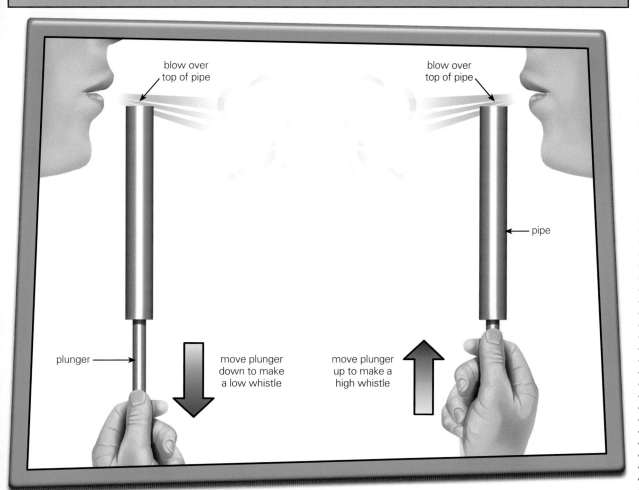

✹ **As the volume of air in the pipe changes as a result of the plunger moving, so does the pitch of the sound.**

A VISUAL DEMONSTRATION OF SOUND WAVES.

What you will need:
- A tuning fork

What you will need to find:
- A small plastic bowl of water and a sheet of paper

Here's what to do:
- Hold the tuning fork by its handle and tap it gently on any firm, but not hard, surface.
- Bring the vibrating tuning fork up to the edge of the plastic bowl and gently touch the bowl.
- Watch the water surface.

What do you observe?

- In this activity, small waves will appear on the water's surface.
- Sound travels through air, as well as through solids and liquids, as a compression wave.
- Hold the sheet of paper in your fingertips.
- Again, tap the tuning fork on a firm surface and bring the vibrating tuning fork up to touch the suspended sheet of paper.
- The paper will flutter as waves of sound are transmitted through it.

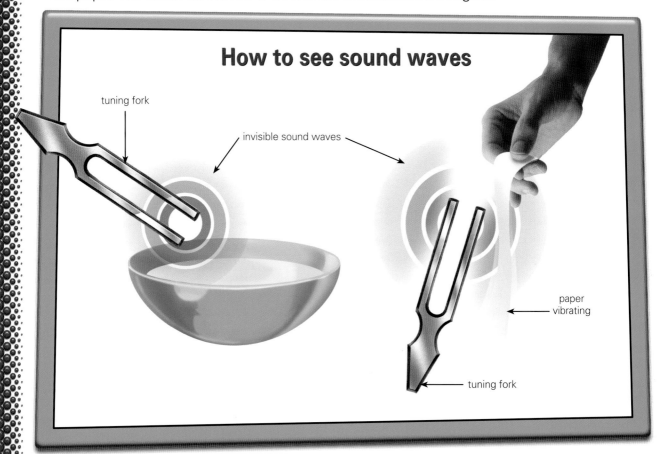

How to see sound waves

EXPERIMENT 14
Vibration of Reeds
General Science

DEMONSTRATE THE SOUNDS CREATED BY VIBRATING REEDS.

What you will need:
- 3 drinking straws

Here's what to do:
- Take any one straw and squash one end of the straw flat.
- Cut that end of the straw into a point as shown in the diagram.
- Place the cut end in your mouth far enough so that you are able to place your lips around the round parts of the straw. This simple instrument is known as a kazoo.
- Now blow into the straw, hard enough to make a sound.
- You will now have made a simple instrument incorporating a vibrating reed.
- While you are making a sound, try to touch the cut end of the straw with your tongue.

What do you feel?

- Repeat the procedure with another straw, this time cutting the straw shorter.

What effect does this have on the sound?

- Use another straw. Cut a small hole halfway along the straw.
- You now have a simple clarinet.
- Cover the hole with your finger and blow into the clarinet.
- While blowing, lift your finger to open the hole.

What happens?

✱ **The pitch of your kazoo is determined by the length of the air column.**

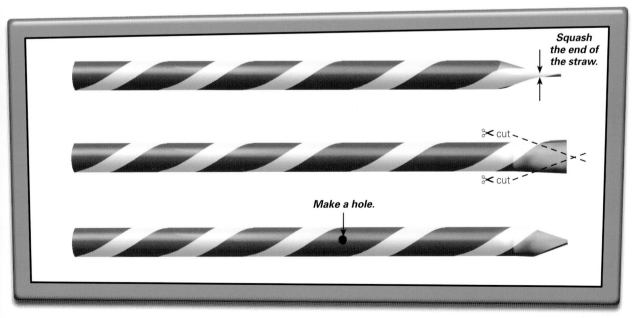

General Science

EXPERIMENT 15
Make a Button Spinner

SPIN A BUTTON FAST ENOUGH TO MAKE A SOUND.

What you will need:
- A large plastic button and a length of cord

Here's what to do:
- The button has two small holes at either side of the center.
- Thread one end of the cord through one of these holes and back through the other.

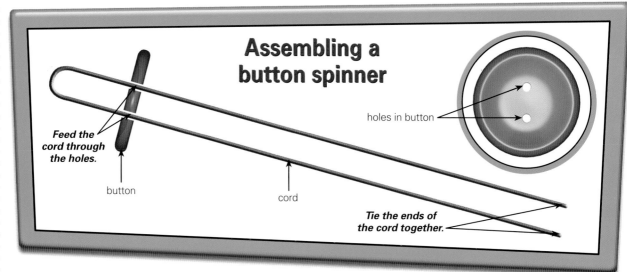

- Tie off the ends so that you have a button on a large loop of cord.
- Hold each end of the loop, with the button in the middle.
- Using both hands, swing the button away from you to twist the cord.
- Pull on the cords to untwist it and relax a little to let it twist a little the other way.
- Quickly repeat this to get the button to spin continuously and quickly.
- It should make a distinct noise.

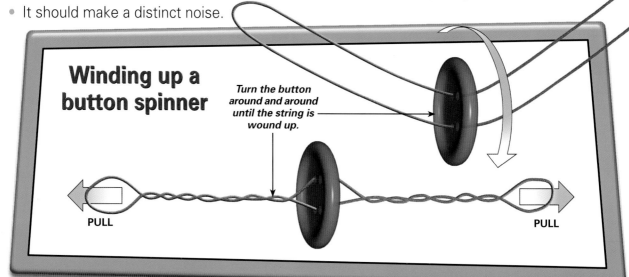

EXPERIMENT 16
Color

General Science

LEARN ABOUT HOW COLORED FILTERS CHANGE HOW WE SEE THINGS.

What you will need:
- The cardboard glasses template and the colored pieces of cellophane

What you will need to find:
- A pair of scissors and some clear tape

Here's what to do:
- Place cellophane of one color over both eye spaces in the pair of glasses and use the tape to attach it in place.
- Notice how things look through the glasses.

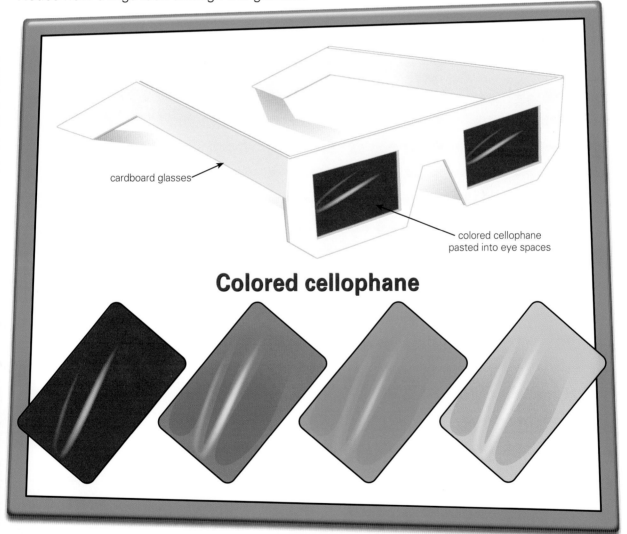

- Remove the cellophane and replace it with a different colored piece of cellophane over the eye spaces.
- Notice how different things look through the glasses.
- ✱ **The colored cellophane is acting like a color filter and changes what your eye tells your brain you are seeing.**

EXPERIMENT 17
Strange Colors

General Science

HOW STARING AT SOME COLORS CAN AFFECT WHAT YOUR EYES SEE NEXT.

What you will need:
- Color dot templates

What you will need to find:
- 4 colored pencils

Here's what to do:
- Color the center circle of each template a different color, e.g., blue, red, green, etc.
- Choose one template.
- Stare strongly at the colored dot on the template for a minute, then quickly look and stare at the white back of the card.

What color do you appear to see?

- Repeat for the other colors.
- Using a library, the Internet, or some other source, find out what you can about complementary colors.

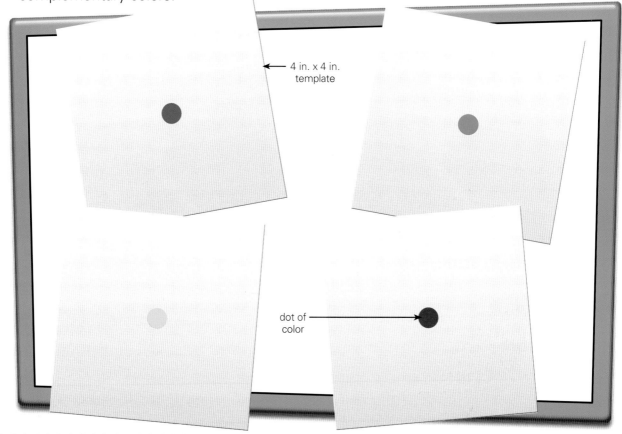

4 in. x 4 in. template

dot of color

EXPERIMENT 18
White Light

General Science

SEE IF THE COLORS OF THE SPECTRUM CAN BE COMBINED TO MAKE WHITE LIGHT.

What you will need:
- The color wheel and a toothpick

What you will need to find:
- A circle of thin, white cardboard; and colored pencils, pens, or paints

Here's what to do:
- Push a toothpick through the center of the color wheel so that you have a spinner.
- Now spin it quickly and see if the wheel seems to go white.
- You probably could not get the wheel to turn white, but you may have seen it become creamy or slightly gray in color.
- As you can see, colored light can combine to form other colors.
- Using a circle of thin, white cardboard and your colored pencils, pens, or paints, make more spinners. Use a different number of colors on each wheel and find out what they combine to make.

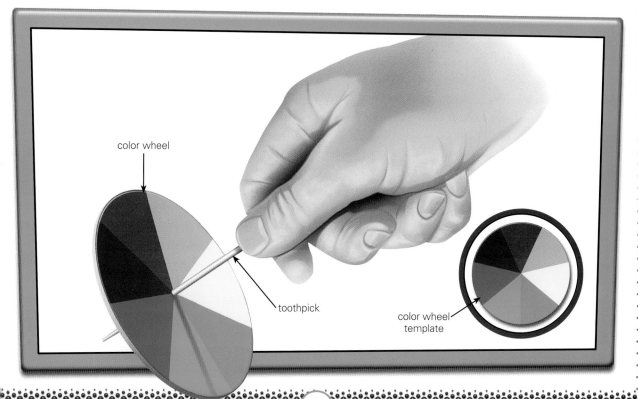

EXPERIMENT 19
A Beautiful World

General Science

USE A PRISM TO SPLIT WHITE LIGHT INTO THE SPECTRUM OF COLORS THAT IT IS MADE OF.

What you will need:
- 1 prism

Here's what to do:
- Close one eye and bring the prism up close to your open eye.
- Look through the prism at your surroundings.
  DO NOT LOOK AT THE SUN OR ANY VERY STRONG LIGHTS.
- You should notice that the objects that you see have colored edges.
- These colors are the colors of the spectrum that make up white light. The colors are red, orange, yellow, green, blue, indigo, and violet.

Some of these will not be clearly visible to you. What colors show up strongly?

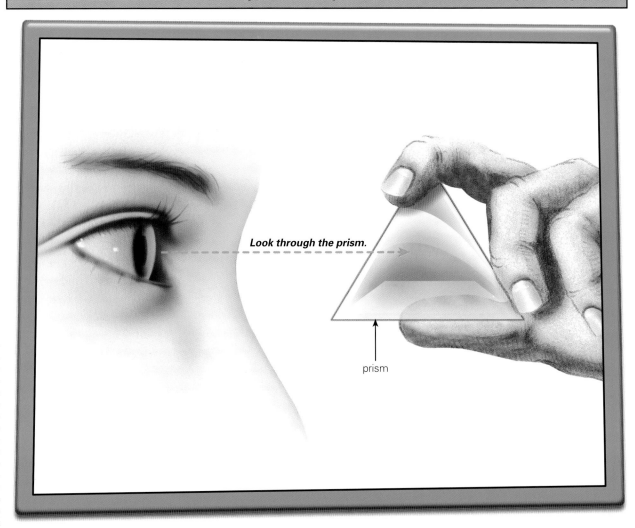

Look through the prism.

prism

EXPERIMENT 20
Making Rainbows

General Science

SEE HOW WHITE LIGHT IS MADE OUT OF DIFFERENT COLORS—THE COLORS OF THE RAINBOW.

What you will need:
- 1 prism

What you will need to find:
- 1 piece of white cardboard and sunlight or a flashlight

Here's what to do:
- Hold your prism up to a light source like the sunlight coming through the window, or use a flashlight.
- Direct the light toward the white cardboard.
- You should be able to see the colors of the rainbow on the white cardboard.

- *Remember, never look straight at the sun.*
- *The prism is refracting (bending) the light, splitting it into the spectrum of colors.*
- *Light is refracted through water droplets in the sky to make rainbows.*

EXPERIMENT 21
Reflection

A DEMONSTRATION OF MIRROR IMAGES.

What you will need:
- Reflection activity templates and a mirror

Here's what to do:
- Place your mirror on its edge along the mirror lines under the words and pictures on the activity sheet, as shown in the diagram.

- You should be able to read the words and see that the pictures are reversed.

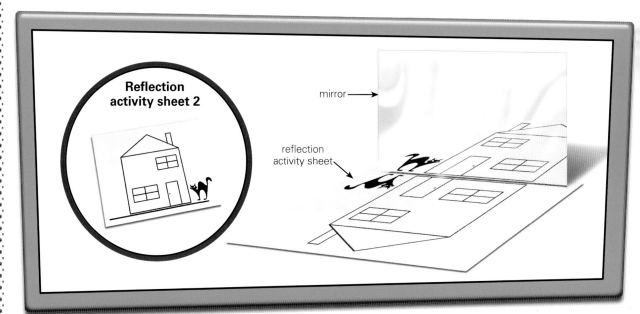

- Mirror images are exact copies, but in reverse.

EXPERIMENT 22
Making a Periscope

General Science

USE THE PRINCIPLES FROM THE PREVIOUS EXPERIMENT TO LOOK OVER A WALL.

What you will need:
- 2 mirrors

What you will need to find:
- A milk carton and a pair of scissors

Here's what to do:
- Cut two windows in your carton and slits on either side of it in the four places shown.
- Place mirrors into the slits with the reflective sides facing each viewing hole.
- Look through viewing hole A and direct viewing hole B at an object. You should be able to see it.
- Try looking over a wall or fence.
- Light is being reflected onto the mirrors and to your eye.
- See what you can see above, below, or around.

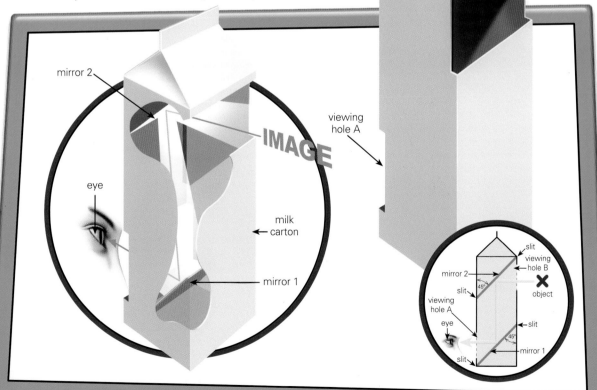

✱ *Don't upset the neighbors.*

EXPERIMENT 23
The Kaleidoscope

General Science

MORE TO DO! MORE TO SEE!

What you will need:
- 3 narrow mirrors

What you will need to find:
- Black paper, waxed paper, leftover pieces of colored cellophane from your glasses experiment, and tape

Here's what to do:
- Tape the three mirrors together (**A**).
- Form them into a prism shape and tape the edges (mirrors on the inside).

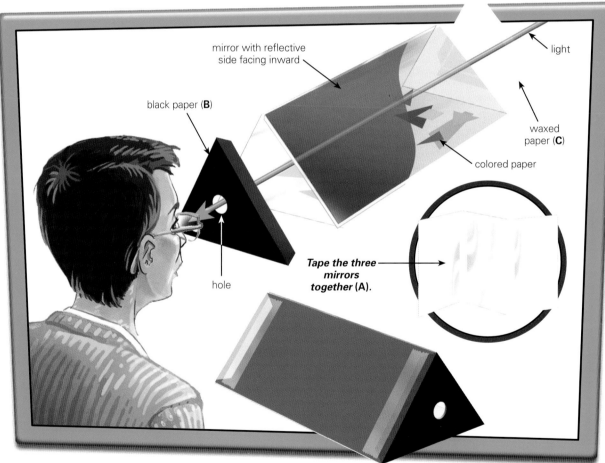

- Place black paper over one end (**B**). Tape it in place.
- Put 10-12 tiny pieces of colored cellophane inside the kaleidoscope.
- Place waxed paper over the open end (**C**). Tape it in place.
- Put a small hole (use the end of a pen or pencil) in the black paper.
- Hold end **C** up to the light from a window (not directly at the sun) and look through the hole.
- By turning the kaleidoscope while looking through it, you will see different effects and patterns.

EXPERIMENT 24
Chromatography

General Science

SEE HOW COLORS CAN BE SPLIT.

What you will need:
- A plastic tube and 3 water-soluble colored pens

What you will need to find:
- A bowl of water, extra colored pens (water soluble), tape, and 10 paper-towel strips (1 in. wide and 3 in. long)

Here's what to do:
- Mark three paper strips with your three colored pens, as shown.

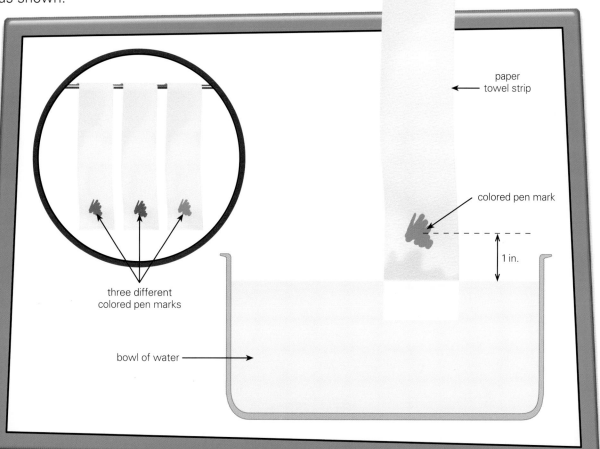

- Tape the strips to the plastic tube, making sure they are not touching.
- Hang the ends in water for at least 5 minutes.

What happens to the colors?

- Record your findings.
- If you have found extra colored pens, try these colors and record them as well.
- Each color's solubility in water is different, so as the water moves up the paper each color is carried a different distance.
- You can see what colors are used to make other colors.

EXPERIMENT 25
Shadows

HOW SHADOWS ARE FORMED WHEN LIGHT IS BLOCKED.

What you will need:
- Shadow templates and the plastic tube

What you will need to find:
- Adhesive tape and additional plastic tubes (pen tubes will work)

Here's what to do:
- Press out the shadow shapes and use tape to attach each one to a plastic tube.
- Use a desk lamp or other light source to make shadows on the wall.
- Make your own shapes and put on a shadow puppet show.

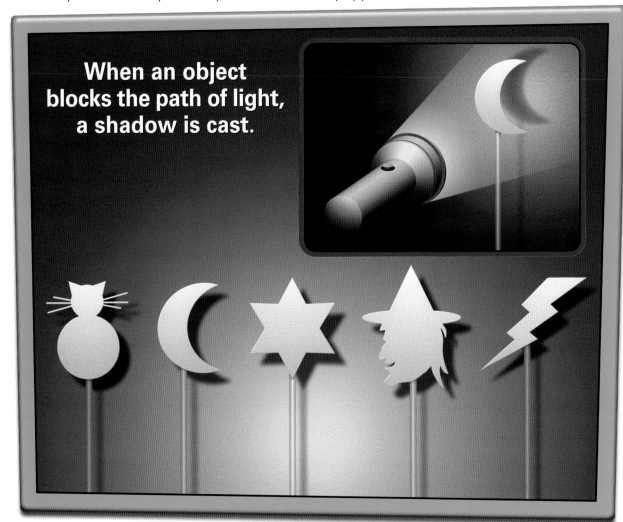

When an object blocks the path of light, a shadow is cast.

* Light travels in straight lines, which is why we see shadows where the light can't go.
* Shadows produced by the sun change as the sun moves. This is how shadow clocks can tell us the time.

ACTIVITY 1
Repulsion Soccer

General Science

PLAY A FUN GAME USING MAGNETS.

What you will need:
- 3 ring magnets

What you will need to find:
- An empty cereal box and a pencil or pen

Here's what to do:
- Cut the cereal box down one side so that you can open it out to as large a piece of cardboard as possible.
- Trim off any pieces that stick out so that you are left with a large rectangle.
- Mark out a soccer field with a pencil or pen. The goals can be marked in the same way.
- Place the three ring magnets on the soccer field so that they repel each other.
- You and a friend each have one magnet, and the remaining ring magnet is the "ball."
- Each of you, as players, must use the repulsion of your ring magnet to move the third ring magnet. Try to score goals in this way while defending your goal.

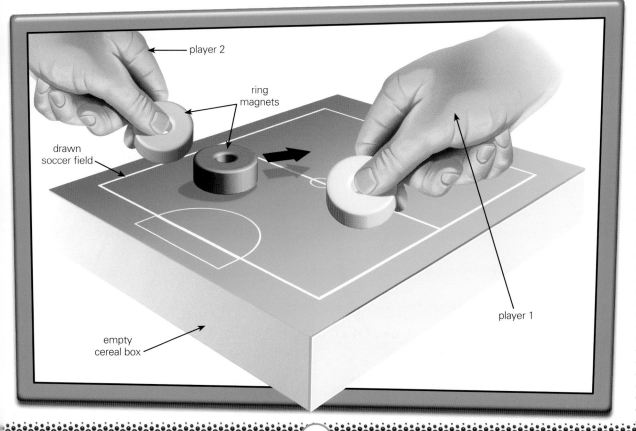

ACTIVITY 2
Boat Races

General Science

USE MAGNETIC ATTRACTION TO MOVE A BOAT.

What you will need:
- 1 bar magnet and paper clips

What you will need to find:
- A shallow plastic tray, some corks, triangles of colored paper (sails), water, and modeling clay

Here's what to do:
- Make each of the corks into a boat using a paper clip to make the mast and keel. Use the diagram to help you.
- Make three islands from the modeling clay and place them in the plastic tray.
- Support the tray on several books so that it is raised off the floor or table enough for you to fit your hand underneath.
- Fill the tray with water until the boats float freely.
- Steer your boats around the tray by guiding them with the magnet underneath the tray.

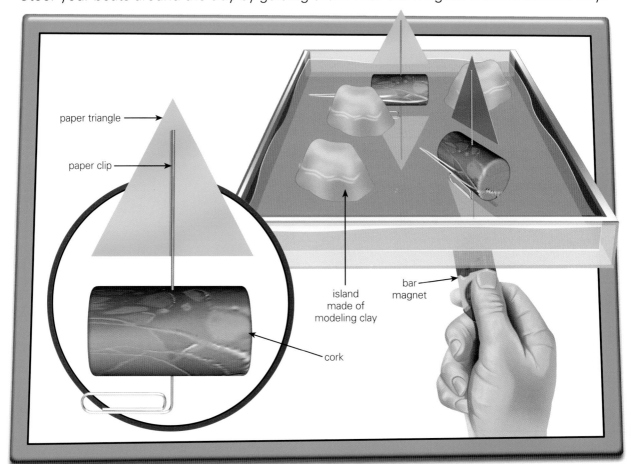

Activity 3
Paper Maze

General Science

USING THE FACT THAT MAGNETS CAN REPEL EACH OTHER.

What you will need:
- 1 bar magnet and 1 ring magnet

What you will need to find:
- An empty cereal box, a pencil or something to draw with, paper or thin cardboard strips, and adhesive tape

Here's what to do:
- Using a cereal box and strips of paper or thin cardboard, make up a maze as shown in the diagram.
- Place a ring magnet in the center of the maze. Use your bar magnet to repel the ring magnet through the maze and move it toward the finish.
- Time yourself for this task.
- Allow a friend to try his hand at moving the magnet through the maze.

Who completed the maze in the shortest time?

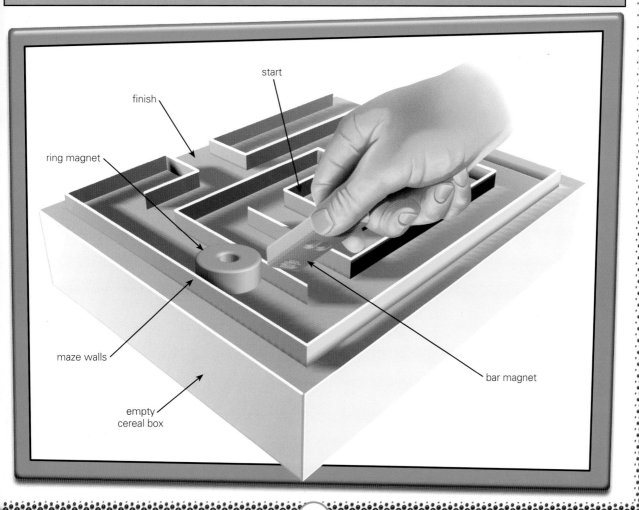

ACTIVITY 4
Pirate Treasure

General Science

USING THE FACT THAT MAGNETS CAN REPEL EACH OTHER.

What you will need:
- 1 ring magnet, plastic rod, and paper clips

What you will need to find:
- Small cardboard box, some old newspaper, cardboard circles, pencils, and string

Here's what to do:
- Tear up the newspaper into small pieces, enough to fill the small cardboard box.
- Prepare some treasure. Write different values like $50, $100, $200, etc., on the cardboard circles.
- Attach a paper clip to each circle. Hide them in the cardboard box of newspaper cuttings.
- Tie the ring magnet to the plastic rod with the string.
- The object of the game is for up to four people taking turns to compete for 30 seconds and collect as much treasure as possible using a magnet on a string to fish for the hidden treasure.

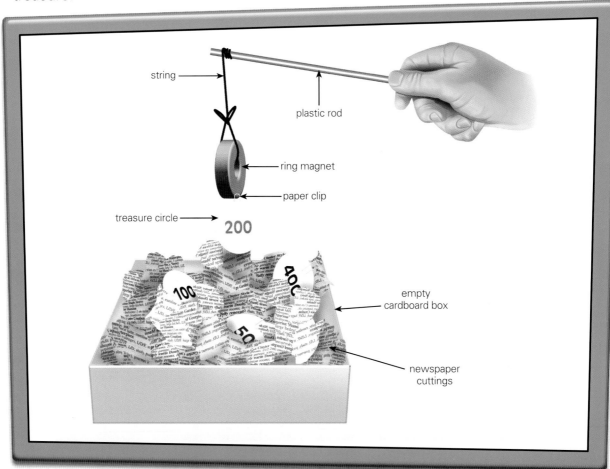

ACTIVITY 5
Cattle Roundup

General Science

USE THE ATTRACTION BETWEEN A MAGNET AND A PAPER CLIP.

What you will need:
- 1 bar magnet and 1 paper clip

What you will need to find:
- A shoe box or a similar box without a lid and some corn flakes

Here's what to do:
- Turn the shoe box upside down so that you have a base under which you can use the magnet.
- At one end of the box, cut out a small hole that will allow the corn flakes to fall through.
- Form the paper clip into a *V* shape and put it on the shoe box. It will be your "cattle dog."
- Sprinkle some corn flakes around the top of the box. These will be your "cattle." Using the magnet underneath the box and the paper clips on top, see how many cattle you can round up in 30 seconds by making them fall through the hole.
- If any cattle fall off, then they are out of the game.
- Do not touch the corn flakes with your hands once you have started to round them up.

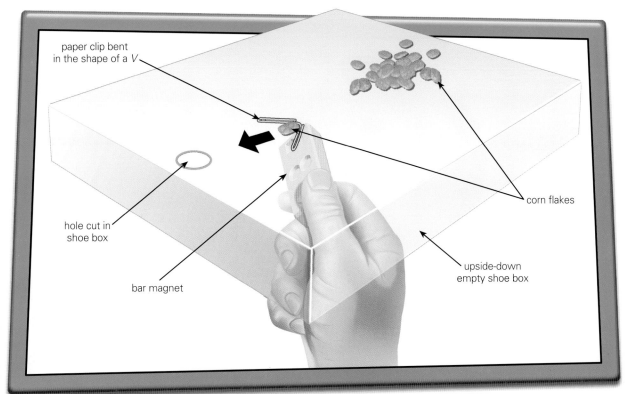

DEMONSTRATE THE TRANSMISSION OF SOUND.

What you will need to find:
- 2 metal spoons and a length of string

Here's what to do:
- Carefully tie one end of the string to the handle of one spoon.
- Suspend the spoon in midair and then tap the spoon with the second spoon.
- Bring the end of the string near your ear and listen.
- Repeat this once more, but this time gently place the end of the string as close to your ear as you can. Tap the spoons together.

What difference do you notice?

- The waves are now transmitted through the string to your ear.
- Solids transmit sounds better than gases.

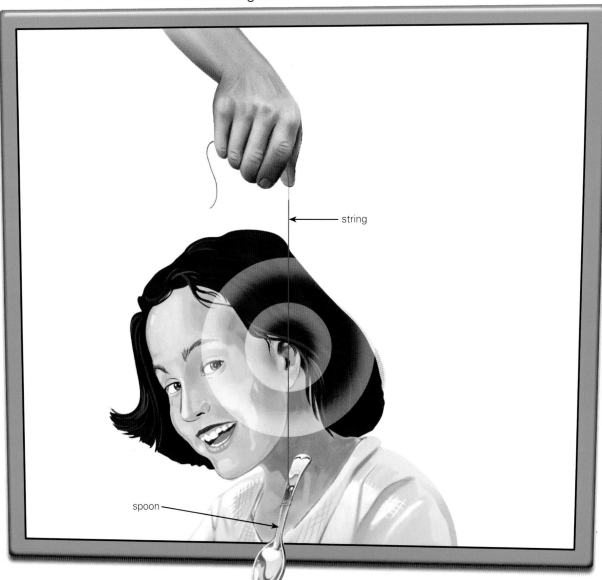

ACTIVITY 7
Communication

General Science

USE VIBRATIONS TO COMMUNICATE

What you will need to find:

- Adhesive tape, 2 empty tin cans or plastic cups, 10 to 13 ft. of string, and 2 small pieces of wood

Here's what to do:

- Put a small hole in the bottom of each can or cup.
- Thread the string through each hole and tie to the small pieces of wood. This is to keep the string from falling out.
- Take one can or cup, give the other to a friend, and stretch the string between you.
- With the string taut, take turns talking into your cup or can while your friend puts the open end of the other cup or can over their ear to listen.
- These old-fashioned toys work because as you talk the vibrations of your voice are passed down the taut string to the other end.

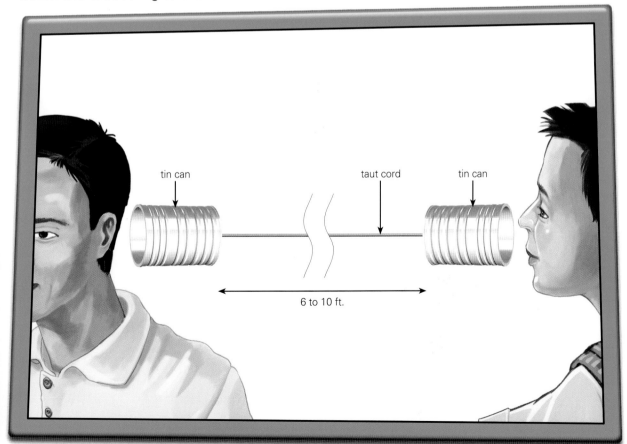

✱ **If using cans, make sure that the open end will not cut anyone.**

EXTENSION ACTIVITY

ACTIVITY 8
Variations on a Button Spinner

General Science

SPIN A BUTTON FAST ENOUGH TO MAKE A SOUND.

What you will need:
- The cord from Experiment 15

What you will need to find:
- Cardboard of various thicknesses and weights

Here's what to do:
- Make several shapes from the different types of cardboard that you have found, as seen below.
- Put holes in them, similar to those of the button used in Experiment 15.
- Thread a piece of cord through each shape as you did with the button.
- Tie off the ends so that each shape is on a large loop of cord.
- Hold each end of the loop with the shape in the middle.
- Swing the shape toward you to twist the cord.
- Pull on the cords to untwist it and relax a little to let it twist a little the other way.
- Quickly repeat this to get the shape to spin continuously and quickly.

Is there any connection between the type of cardboard or its weight and the sound it makes?

- Try some different shapes and different numbers and sizes of holes.

Does changing the edges of the shapes have any effect?

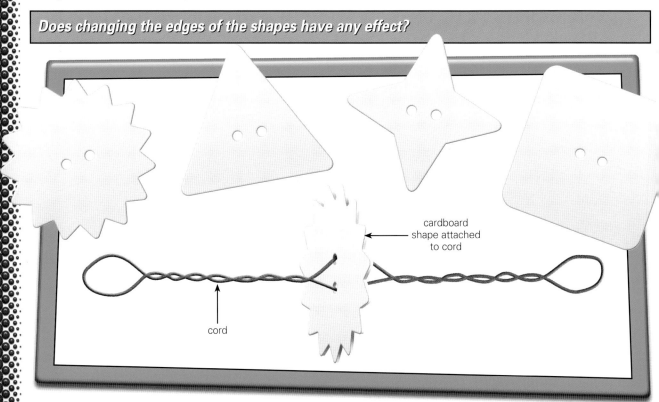

cardboard shape attached to cord

cord

✷ *You could try putting points on the edges or gluing on small pieces of squashed straws.*

Activity 9
Amplitude and Frequency

General Science

HOW AMPLITUDE AND FREQUENCY AFFECT SOUND.

What you will need to find:

- Strip of wood or a ruler

Here's what to do:

- Place the strip of wood on a table with at least two thirds of it hanging over the edge.
- Hold the short end firmly to the table.
- Firmly flick the free end down.
- Note the sound made.
- Repeat with less of the wood over the edge of the table.
- Note the sound made.
- Repeat once more with only a short amount of wood over the edge of the table.
- Again, note the sound made.

Did the sound become higher or deeper as the free end of the wood became shorter?

- The difference in the sound is due to the change in pitch.
- As the free end of the wood becomes shorter, it vibrates faster.

Therefore, as the frequency increases, does the pitch become higher or lower?

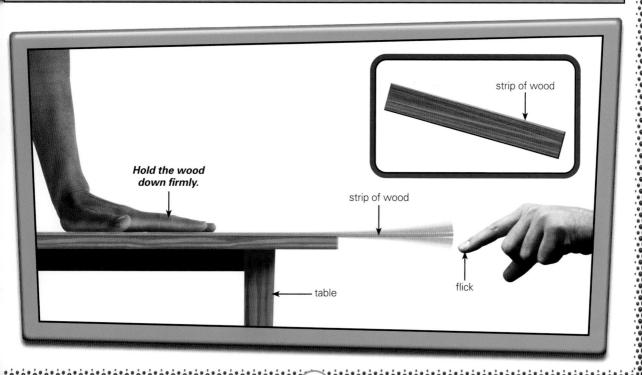

41

EXTENSION ACTIVITY

General Science

LEARN SOME MORE THINGS ABOUT A PRISM.

What you will need:
- 1 prism

What you will need to find:
- A piece of blank cardboard and a flashlight

Here's what to do:
- Write the word "PRISM" in capital letters on the piece of blank cardboard.
- Place a face of the prism over the word and look at it from the front until you see an image of the word in the prism.

Is the image the same as or different from the original?

- Now put the prism directly over the word "PRISM" and look at the prism from above.

How does the image of the word look this time?

- Close one eye and hold the prism close to your open eye. Move the prism around until you can see your feet in the prism.

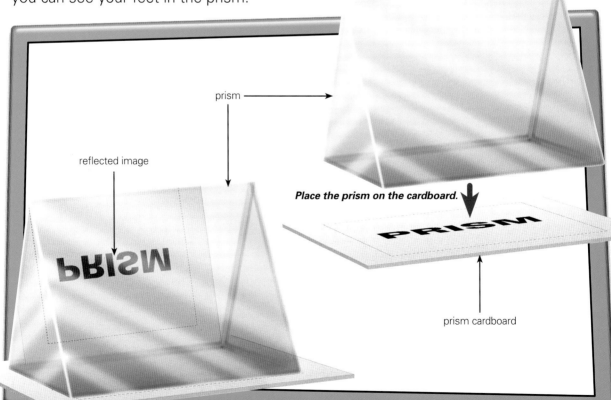

✱ *A prism can act like a mirror.*

ACTIVITY 11
Straight Lines of Light

General Science

A DEMONSTRATION SHOWING THAT LIGHT TRAVELS IN STRAIGHT LINES.

What you will need:
- 1 mirror

What you will need to find:
- 1 flashlight, 2 cardboard tubes, and modeling clay

Here's what to do:
- Stick your mirror onto a wall with the modeling clay.
- Get someone to shine a flashlight down one tube as shown.
- Look through the other tube like you would a telescope and try to see the light.

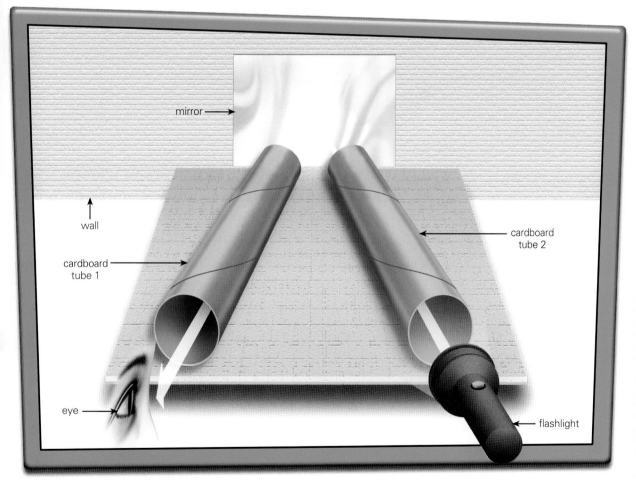

- **The light reflects off the mirror and down the tube to your eye, depending on the angle (direction) of the flashlight.**
- **Light travels in straight lines.**

EXTENSION ACTIVITY

ACTIVITY 12
Strange Chromatography

General Science

SEE HOW COLORS CAN BE SPLIT.

What you will need:
- A plastic tube and 3 water-soluble colored pens (black, green, brown)

What you will need to find:
- A bowl of water, extra colored pens (water soluble), paper towels, and adhesive tape

Here's what to do:
- Cut your paper towels into strips like the ones from Experiment 24.
- Using several water-soluble colored pens, draw a simple face and other pictures on each strip.
- Set your strips up to separate colors as in Experiment 24.
- Leave for 5 minutes and look at your results. They should be very interesting.
- Your strips can be left to dry and then used for a simple bookmark or used in other interesting ways.

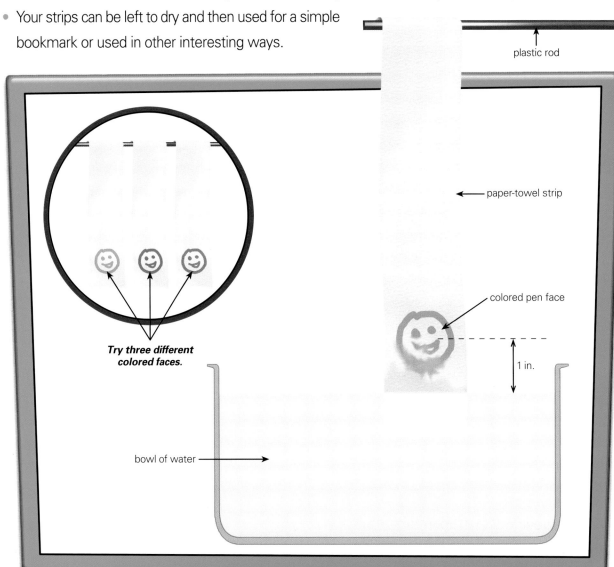

Try three different colored faces.

ACTIVITY 13
Seeing Around Corners

General Science

REFLECTING LIGHT OFF OBJECTS AND ONTO MIRRORS CAN CARRY IMAGES FROM AROUND CORNERS TO YOUR EYE.

What you will need:
- 4 mirrors

What you will need to find:
- A shoe box or cereal box and modeling clay

Here's what to do:
- On a table, set up the box and mirrors as shown using modeling clay to hold the mirrors in place.

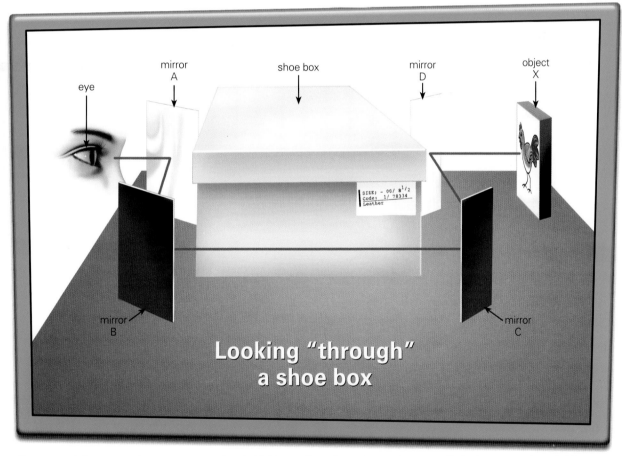

Looking "through" a shoe box

- Put an object on the spot marked X. By adjusting the mirrors slightly, you should be able to see it without looking over the box.
- Get someone to change the object without you seeing and see if you can identify the object.
- You are seeing around corners.
- The light reflecting off the object and onto the first mirror is then reflected to the second, third, and fourth mirrors and then to your eye.

ACTIVITY 14
More Reflection

MORE TO DO! MORE TO LEARN!

What you will need:
- 1 flexible safety mirror, the "wire" template, and the "house" template

What you will need to find:
- 1 sheet of paper, pens or pencils, and a friend or parent to help you

Here's what to do:
- Using your mirror and the "wire" mirror template, see if you can write backward or upside down.
- Draw letters like those below while watching in your mirror.
- Now with the help of a friend, have them bend the mirror so you have a slightly convex curve facing you, reflecting the image of the "house" template.

Can you copy the writing?
What has happened to the writing?

- Now have your friend slightly bend the mirror to produce a slightly concave shape.

What has happened to the reflection of the writing?

- Now try gently twisting the mirror.

Now what has happened to the reflection?

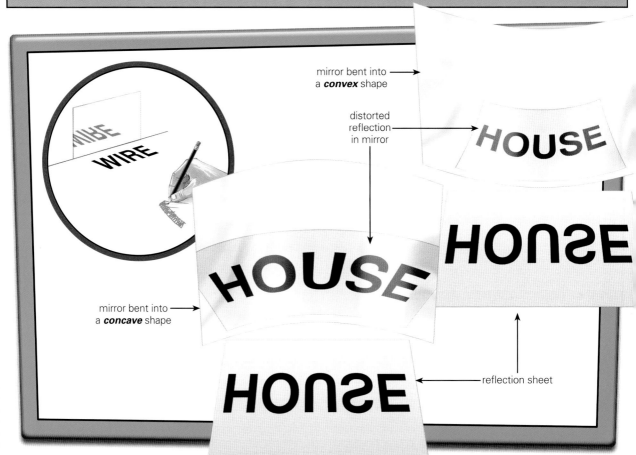

ACTIVITY 15
Multiple Images

General Science

SHOW THAT TWO MIRRORS AT AN ANGLE TO EACH OTHER CAN PRODUCE MULTIPLE REFLECTIONS.

What you will need:
- 2 mirrors

What you will need to find:
- 1 toothpick and modeling clay

Here's what to do:
- Put a small piece of modeling clay onto the end of the toothpick so that it can stand upright by itself.
- Put modeling clay on the bottom edges of each mirror so that they can stand upright.
- Place the mirrors end to end in front of you so that they are in a straight line. Stand the toothpick about 2 in. in front of them.

How many images of the toothpick do you see?

- Now angle the mirrors in toward each other a little.

How many images of the toothpick do you see now?

- Repeat using increasingly sharper angles between the mirrors.

What do you notice about the number of images?

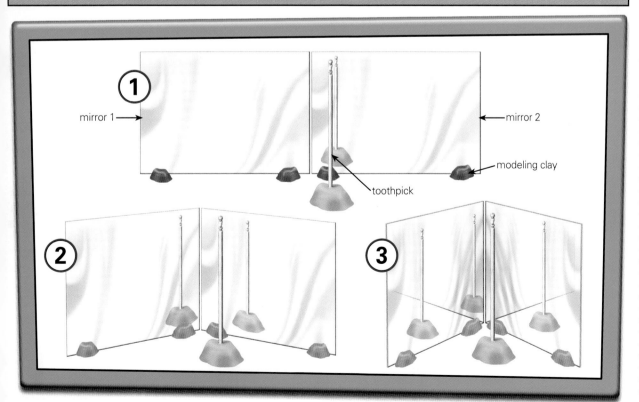

✶ *Kaleidoscopes use angled mirrors to make multiple images.*

EXTENSION ACTIVITY

ACTIVITY 16
Mirror Image Painting

General Science

USING THE FACT THAT MAGNETS CAN ATTRACT SOME METALS.

What you will need:
- 1 bar magnet and 1 paper clip

What you will need to find:
- A piece of white cardboard and some nontoxic poster paints

Here's what to do:
- Fold a piece of white cardboard, then unfold it and lay it flat.
- Place two small amounts of poster paint on one side of the crease line.
- Place a paper clip on the poster paint.
- Use the bar magnet from underneath the cardboard to attract the paper clip and randomly move it through the paint.
- Remove the paper clip and carefully fold the cardboard together again, making sure to press lightly.
- Unfold the card to reveal the mirror image of the abstract art created on the other half.
- Leave to dry.

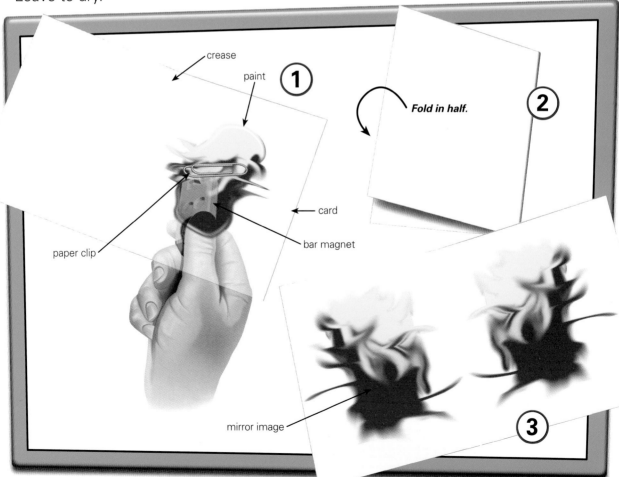

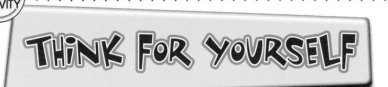

General Science

Here are nine experiments for you to try. We have given you a basic background for each one. Try to complete them using the knowledge you have gained from the previous experiments.

17. Cleaning an Aquarium

What you will need:
- 1 bar magnet, 1 ring magnet, cotton rag, 1 large glass or jar, 1 candle, and water

Here's what to do:
- Ask an adult to blacken the inside of the glass with candle smoke. Then fill the glass with water.
- Wrap the ring magnet in the cotton rag and place it in the glass.
- Use the bar magnet to attract the ring magnet to the side of the glass.
- Move the magnet around the inside glass surface to remove the smoke marks.

18. Silent Signals

What you will need:
- 1 mirror and a sunny day

Here's what to do:
- Position the mirror in such a manner that you can reflect the sun's light rays onto a near object.
- Now carefully reflect the light to a more distant object.
- ✸ **By flashing the light for longer and shorter intervals, it is possible to signal messages over long distances (up to 2 miles with a handheld mirror). This method of signaling is known as heliographs. In 1886 General Nelson Miles used heliographs in his battles with Native Americans in Arizona.**
- ✸ *Signaling mirrors are often standard equipment for explorers and are often found in life rafts as safety equipment.*
- ✸ *See if you can locate a copy of Morse code and then learn to signal a friend.*

19. Use Shadows to Tell the Time

What you will need:
- 1 paper plate, a straight stick or pencil, and a pencil or pen to write with

Here's what to do:
- On the rim of the paper plate, write *N*. Make a small hole in the center of the plate.
- Put the plate outside, on a flat surface that is in full sun for most of the day.
- Make sure that the *N* faces north. You could use a compass or ask someone who knows.
- Push a pencil or stick into the hole in the center of the plate and hold it in place with modeling clay. Make sure that it stands up vertically. Every half hour, mark where the shadow is on the plate. Also write down the time.
- Write the month on the plate.
- You now have a simple sundial for a particular month.
- In a more sophisticated sundial, the pencil or stick is replaced with a triangular piece called a gnomon. This points to the North or South Pole, depending on which hemisphere you live in. The angle that the gnomon makes with the horizontal is the same as your latitude.

General Science

20. Didgeridoo

What you will need:
- Long cardboard tube such as from an old roll of aluminum foil or paper towels, paints, and brushes.

Here's what to do:
- A didgeridoo is a hollow tube into which you place your lips so that the sides of the tube fit firmly around your mouth.
- This stops air from coming out when you blow into the tube.
- You blow down the tube with your lips very loose so that they vibrate, almost like blowing a "raspberry."
- You are trying to produce a deep sound.
- Practice with your tube, and when you are able to produce interesting sounds, decorate it with the paints.
- Try using tubes of other sizes, especially long ones.
- If the tube is too wide, you may have to find a way of reducing the size of the opening so that your mouth will fit. Some people use softened wax.
- Get some friends to make their own didgeridoos and play a tune together.

21. Ancient Musical Instrument

What you will need:
- Internet access

Here's what to do:
- The bullroarer is a sound-making instrument used by Australian Aborigines. They also use other instruments to make music.
- Search for information on other musical instruments used by Australian Aborigines and other cultures and learn about how the sounds are made with each instrument.
- Some instruments that you can look for are the didgeridoo, kalimba, shakuhachi, and ocarina.

22. Use Your Knowledge of Reflection

What you will need:
- 1 mirror from this kit, pens or pencils, and 1 sheet of paper

Here's what to do:
- Put the mirror next to a page from a book or newspaper.
- See if you can write in reverse by copying what you see in the mirror.
- Write a word in capitals or block letters on your sheet of paper.
- Without using a mirror, try to write the word in reverse.
- Use your mirror to look at the original word and compare your effort at writing in reverse to what you see in the mirror.

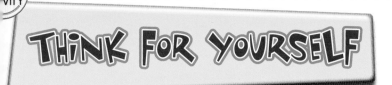

General Science

23. Even More Reflections

What you will need:
- 1 plastic mirror, paper, and pencils

Here's what to do:
- Bend the mirror slightly and hold it next to some writing or an object.

> *What do you notice about the reflection in the mirror?*

- Repeat, but this time bend the mirror a little more. Don't break it.

> *What effect does the extra bending have?*

- Draw a cartoon character on some paper.
- Bend the mirror slightly around it and draw what your cartoon character looks like.

> *As the reflection of your character goes further into the mirror, how clear is the image?*

24. Is This a Kaleidoscope?

What you will need:
- Your kaleidoscope from Experiment 23.

Here's what to do:
- Remove the waxed paper from **C** and remove the colored pieces from inside.
- You now have what is called a teleidoscope.
- Look at your surroundings through the hole in the black paper.
- This close cousin of the kaleidoscope makes everything a little more interesting to look at.

25. Sounds Travel Better in Denser Air

What you will need:
- A watch or clock that "ticks" and a large balloon to blow up to about 12 in. in diameter.

Here's what to do:
- Hold the ticking watch or clock about 12 in. from one of your ears and note the loudness of the sound.
- Now blow up the balloon so that it is about 12 in. in diameter.
- Place the balloon to your ear and lightly place the watch on the other side of the balloon.

> *What difference is there in the sound that you hear?*

- The ticking of the watch is passed through the balloon to the air inside.
- The air vibrates and passes the sound on to the other side of the balloon and to your ear.
- Since you have blown air into the balloon under pressure, the air particles are squashed in and closer together. This means they can vibrate into each other more quickly than outside the balloon, so sound moves faster.
- Sounds travel even faster in denser materials like liquids and solids.
- Put your ear to a metal pipe and tap it at an arm's length from you. The sound should be much clearer and louder than if you were just standing near the pipe.

✻ *Be sensible and don't break the balloon near your ear, as it could cause pain and damage your ear.*

General Science

THE FOLLOWING IS A LIST OF SOME OF THE MORE DIFFICULT WORDS IN THIS WORKBOOK AND WHAT THEY MEAN.

Align
Describes how iron filings change their position (line up) when under the influence of a magnet. They align themselves within the magnet's magnetic field.

Circuit
A circuit is considered to be a complete loop. In the case of an electrical circuit the loop is completed when the switch is moved to the ON position. In this book we refer to different types of circuits:
  Series Circuits
    Components are connected from end to end. If one component is removed or disconnected, the circuit is broken.
  Parallel Circuits
    Each component is connected separately. If one of the components is disconnected, it does not affect those with which it is parallel.

Classify
To sort objects into common groups. For example, we may classify materials as magnetic (such as steel and nickel) or nonmagnetic (such as rubber and plastic) materials.

Complementary colors
Are colors that are opposite each other on the color wheel. Violet and yellow are complementary, as are blue and orange and red and green.

Compression wave
Compression waves are a particular back-and-forth motion that creates compressed or longitudinal waves. It appears as if something is actually moving along the material, but it's really just the distortion moving, with one part influencing the next.

Frequency
The number of times any event is repeated in a period of time. Sound waves have a frequency, which is the number of pulses that go past a fixed point in a given amount of time. You hear sounds when the pulses of compressed air excite your eardrum, which sends signals to your brain.

Image
A copy or likeness of something. For example, a mirror may reflect your image. You can even have an image in your mind when you remember something vividly.

Magnetic field
The area near a magnet where it is possible to detect the lines of magnetic force. The lines of magnetic force surrounding a bar magnet can be demonstrated by placing the magnet over a bed of iron filings. The iron filings will align themselves in a pattern that shows the lines of magnetic force.

Magnetized
Some metals show no signs of magnetism, but when they are stroked with another strong permanent magnet, they become magnetized. They now act as a weak permanent magnet.

Permanent
Refers to a magnet with an ongoing magnetic effect. An electromagnet only has a magnetic effect while an electric current flows through its coil.

SAY WHAT?
Understanding Difficult Words

General Science

Pitch
The quality of a tone or sound as determined by the frequency of the sound vibration. The greater the frequency of vibration, the higher the pitch.

Primary colors
There are three primary colors: red, blue, and yellow. The three primary colors in science when studying light are: red, blue, and green.

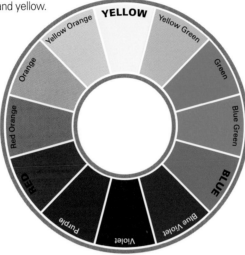

Primary colors can be mixed to create all other colors. You can mix two primaries to get a secondary color. You will notice that each secondary color on the color wheel is bounded by two primaries. These are the components that you would mix to get that secondary color. There are three secondary colors: green, orange, and violet.

Reflect
To cause a change in direction. In the case of a mirror, the image you see in the mirror is called a reflection. Light from a flashlight is reflected outward by the reflective surface of the flashlight's bulb.

Repel
To force away. Magnets will repel each other if the north pole of one magnet is brought near the north pole of another. If unlike poles are brought together, the magnets attract each other.

Soluble
A material is soluble if it can be dissolved in another material. For example, salt may be dissolved in water at room temperature until the water becomes saturated and any more salt that is added remains undissolved.

Spectrum
The color spectrum is a series of colored bands into which white light is broken when it passes through a prism. A rainbow is an example of white light being broken into the colors violet, indigo, blue, green, yellow, orange, and red.

Template
A template is a prepared shape or design. Templates are used to ensure that accuracy can be repeated. In this book, several templates are used to make sure that items are accurate.

Temporary
Lasting for a set time. Its use in this book refers to an electromagnet with a temporary magnetic effect, lasting only while an electric current passes through its core.

Transmit
To pass along or send out. For example, sound may be transmitted through water, air, or along solid objects. Sound waves transmitted through the air or along a metal rod or a thin piece of paper may be vibrated by the sound waves.

Notes

General Science

Notes

General Science

Notes

General Science